An Addendum At The Beginning!

When writing this book I must tell!
I was put through………Technology Hell!
    When the computer crashed,
     All my hopes were dashed,
As I thought to myself………..Oh, Well!

      But then the Apple Techie told me with glee!
      "A file suddenly appeared………...magically!"
        So to print it did go,
        But I did not yet know!
      That 'Computer Madness' had not set me free!

          Because the file that was recovered………in fact!
          Was from two edited versions………way back!
            With spelling mistakes,
            And for goodness sakes,
        Craig Johnson, Mark Sandhoefner and Sam Goff………..were all sacked!
              (from prompt ix, last paragraph!)

          So when you're reading this book keep in mind,
          Grammatical perfection………you will not find!
            But enjoy and have fun!
            From the beginning till done!
          Cause its 'Sense of Humor'……… is how it's defined!

OPEN BOOK

"No error message!"
This book is not opening!
You must turn the page!

HECK NO!
TECH NO!

A HUMOROUS GLIMPSE AT THE MADNESS OF TECHNOLOGY!

HECK NO!
TECH NO!

A HUMOROUS GLIMPSE AT THE MADNESS OF TECHNOLOGY!

by

Patricia Studts

Illustrations by Keila Ramos

MINNETONKA, MINNESOTA

COPYRIGHT © 2007 by patricia stutts

All rights reserved, including the right of reproduction in whole or in part in any form, including, mechanical, electronic or digital, without written permission from the Publisher, patafea books, (www.patafeabooks.com); the exception being for brief excerpts quoted for review.

HECK NO!
TECH NO!

A HUMOROUS GLIMPSE AT THE MADNESS OF TECHNOLOGY!

ISBN: 978-0-9800985-0-1

FIRST EDITION: NOVEMBER 2007

Printed in The United States of America
by Sentinel Printing Company, St. Cloud, Minnesota

This book is printed on acid free paper.

Book Design by patricia stutts

DEDICATION

This book it must be dedicated to…….
All Technology users……can you guess who?
    It's the Whole Wide-World!
    For we've all been hurled,
Into a Technology-Madness Zoo!

ACKNOWLEDGEMENTS

I would like to thank the following people who have influenced my life and helped to mold the person who wrote this book...............Moi, of course!

Bertha Woskoff - My Austrian, maternal Nanny, who lived to the young age of 89 and never lost her accent! or her witty and light-hearted sense of humor; which I feel blessed to have inherited!

Max Weissman - My Russian, paternal Grandfather, who was a terrible tease; which, I also inherited! (And here I thought that I had never won a Lottery!) Over the years I had to learn to curb the degree of teasing since family, friends and strangers were never quite sure if I was being serious or....................truly, just teasing!

Mr. Browne - Headmaster of Freeport High, Freeport, Grand Bahama. Given the Brits' love of Poetry, Literature and Limericks, I began my writing journey at a time when The Bahamas were still a British Colony (Ah! but alas, another one lost!) and the school was under the British system, albeit loosely (one might even go so far as to say that it was dangling by a thread)!

All of my living relatives and friends who must have had some influence on who I am today! They know who they are and I am sure they will let you know who they are, especially if you're within earshot of them! Those of them who listened and critiqued the work in progress: Al and Leslie Kenrick, Sean and Sibel Kenrick, Nancy Warren, Paul Wood, Sandford Sacks, Maddy Mayor, Shelah Yeatman, Ken Debono, June and Jere Cleaveland, and Janet Schwartz (Mom!)

Sandy Accola - A friend, whose encouragement, wisdom, kind words and belief in me, gave me the desire and faith to move forward with this project; a project that had been sitting on the back burner..............on simmer, I might add, for a very long time! I recognized during a long conversation with Sandy (back in August) that this book, which didn't have a name yet, was going to become a reality...................and so, as you can see, it has!

Jeremey - An Out-Sourced, Support Techie in, Toronto, Canada. After the Tech Support part of the Tech-Support call, Jeremey, a college student, educated me about, 'TECHNOSIS' and 'TECHNOSTRESS', which he was studying in Psychology and Sociology classes. He suggested I GOGGLE them...........a whole new world of, TECHNOBABBLE, opened up to me and inspired the section of, 'TECHNO' Limericks!

Keila Ramos – Illustrator, with her incredible talent, who brought my doodled, 'KeyBoard Head', character to life, along with all of the other wonderful illustrations in this book!

Diana Green (Academy College), who, after receiving a frantic call from me, told Keila about the project, and as they say..................the rest is history!

Sentinel Printing Company - Everyone there who touched this book and made it happen!

Lastly, I would like to thank those who have influenced me in some special way or were directly instrumental in making this book possible: Josh and Jen Clayton, Robert Schmidt, Gene Moore, Mike Bradshaw, Emory Dively, Anju and Ashu Kataria, Eddie Ulrich, Apple and Kinkos employees who helped with the many Techno Challenges (Ridgedale), Jared Stutts.............and Moi, of course, for believing in myself, even when most everyone else already did! And to those who didn't....................Oh, Well!

PREAMBLE TO THE TABLE OF CONTENTS

Open Book..prompt i

Title Page...prompt iii

Second Title Page!..prompt v

Copyright Page..prompt vi

Dedication ....prompt vii

Acknowledgements..prompt viii

Preamble to The Table of Contents...prompt x

Postamble to The Preamble of The Table of Contents..................prompt xi

TABLE OF CONTENTS

Prolouge	Prompt	1
My Technology Beginnings	Prompt	4
Read Out Loud	Prompt	8
HECK NO! TECH NO! CREDO!	Prompt	10
Poetry by patricia stutts	Prompt	12
'TECHNOBABBLE' Match Game	Prompt	34
Japanese 'TECHNO'	Prompt	42
Haiku From The Web	Prompt	45
Haiku by patricia stutts	Prompt	52
Japanese Smileys	Prompt	68
Smileys and Emoticons	Prompt	72
'TECHNO' Limericks	Prompt	76
Blog	Prompt	104
Technology Theme Party Menu	Prompt	116
Top Ten Lists (Love/Hate About Technology)	Prompt	120
The Birth of the KeyBoard Head	Prompt	124
About The Author	Prompt	127
About The Illustrator	Prompt	128
Order Information	Prompt	129
Close The Book	Prompt	131
patafea books Logo	Prompt	133

PROLOGUE

How did this book come about? Well, it's been a long journey from there to here! To synopsize it, it is a coming together, a marriage you might say, of my humorous nature (my ability to see humor in almost everything) and my somewhat intense dislike for the ever, second-changing world of Technology that we live in today! You know the old addage (one that Technology keeps well and alive!), 'Can't live with it, but can't live without it, either!' And so, I go through each day hoping not to spend more than four hours a day on a tech-suport call somewhere in the world with one of the litany of companies that we are coerced into dealing with in order to navigate through our lives! What happened to the 50s and 60s! Weren't we still in the Industrial Age back then?

Along with my original work, I have added outside Technology humor drawn from several different venues. Before embarking on this book, I had never read a Blog! Had I known how funny they could be, I would have started reading them a long time ago! I have included some fun ones, however, some of my very favorites I had to leave out because then the book would no longer have a 'GP' Rating!

Also, in researching various subjects, I discovered some Technology dictionaries: netlingo.com, UrbanDictionary.com, wikipedia.com. They are full of information and funky words, some of which I have included for you to match up with their definitions..........Good Luck!

PROMPT 1

As Technology is 'WorldWide', both in its applications and with its challenges, I have also included a section called, Japanese 'Techno'. I found a wonderful collection of Japanese Smileys and Haiku Poetry, to which I have added some of my own.

Although it was my intention to first write a couple of childrens' books based on true stories, this book reared up its pages back in January of this year with the poem, 'Computer Madness', a Dr. Suessesque poem! After five hours (the record to date) on a tech-support call, this poem poured out of me...............and thus was the conception of this book. Hmm, let's see, that's almost eleven months of gestation, with the last month being......................hard labor! Maybe next time I should think about having a, C-Section!

So, when you are a Megabyte away from going into a 'Techno Rage' or a Gigabyte away from 'Prompt Insanity', due to one of those iZillion Techology challenges that stare you in the face everyday................grab, 'HECK NO! TECH NO!' and have yourself a hearty Terabyte laugh! Enjoy!

'MY TECHNOLOGY BEGINNINGS'

My love/hate relationship with Technology began with this, Tandy Radio Shack Computer, shown to the right. Who knew then about Technology Stress, Madness and Rage! I remember, as if it were five seconds ago, my first experience with computer rage! I was entering in, ironically enough, my poetry, into what was then, pretty much, a Word Processor. After about an hour of not being able to locate what I had already entered, I calmly walked into the Utility Room, picked up a hammer, walked back into the Sun Room (where the computer was located in our San Francisco Flat), laid the hammer next to the computer, and proceeded to hunt down the files; knowing that if I couldn't locate them in the next ten minutes, the computer would become pieces of 'a work of art' that I would create in the form of a Table Sculpture or Wall Hanging! However, the frantic call that I had made to my husband (before cell phones), was returned just in the nick of time……..and wouldn't you know, I was only one keystroke away from insanity………..and a new art form! And thus began the Computer Illogic, "If you know how to do it, it's simple!" Illogical because, we all know that even if we do know how to perform certain tasks on our computer,………………it is NOT always simple and most definitely NOT logical!

Radio Shack TRS-80 (Model I)
Catalog: 26-1001
Released: August 1977
Price: US $599.95 (with monitor)
How Many: 200,000 (1977-1981)
CPU: Zilog Z-80A, 1.77 MHz
RAM: 4K, 16K max*
Ports: Cassette I/O, video, Expansion connector*
Display: 12-inch monochrome monitor 64 X 16 text
Expansion: External Expansion Interface*
Storage: Cassette storage*
OS: BASIC in ROM*
* Additional ▫nglish▫▫ies with Expansion Interface

PROMPT 5

"Where's the computer? It's in the keyboard! As one of the first home computers ever, the TRS-80 was a great success. Tandy wasn't expecting many sales, but this, their first computer, sold 10,000 units in the first month alone. It includes everything you need to have a real computer of your very own – the computer, monitor and cassette deck for loading and saving data. Yes, these were the days when you bought, loaded and saved your data and programs on cassette tapes. Floppy disk drives didn't come into common use until years later. Even then, they were very expensive, costing hundreds of dollars. The TRS-80 Mini-Disk was available within a year of the TRS-microcomputer's release, but it cost $499, more than the computer itself. Even three years later, in 1980, the TRS-80 floppy drive still costs about $425."

This information was found at, oldcomputers.net.

PROMPT 6

READ OUT LOUD

Before you flip this page to proceed,
Disable the Mute Button, then begin to read!
  For you have the choice
  To use your OWN voice
For a much better experience................Indeed!

PROMPT 8

HECK NO! TECH NO! CREDO!

CHANT

CHANT

CHANT

CHANT

TECH NO! IN
HECK NO! OUT
HECK NO!
TECH NO!
IS WHAT
WE'RE ABOUT!

TECH NO! SPEND
HECK NO! MORE
HECK NO!
TECH NO!
STAY OUT
OF THE STORE!

TECH NO! WHAT?
HECK NO! DOUBT!
HECK NO!
TECH NO!
IS WHAT
WE SHOUT!

CHANT

CHANT

CHANT

CHANT

POETRY

by

Patricia Stutts

HECK NO! TECH NO!

HECK NO! TECH NO!
Is it some kind of dance?
Or perhaps it's simply
A Cyberspace romance!

It's neither of those
Or many things more!
It's Technology Madness
Oozing from every pour!

In a world that's moving
Way too fast,
Today's tomorrow
Is now in the past!

There's more Technology
Than you'll ever need,
And you'll wonder how
You will ever get freed!

But I'm here to state
That it's okay to say
HECK NO! TECH NO!
In your life today!

And while it's obvious
You can't get rid of it all
You certainly can avoid
The TECHNOCREEP Crawl!

Pick and choose carefully
What Technology you buy
So there's time left over
To give, 'Real Life' a try!

HECK NO! TECH NO!
Is the way to proceed!
Then from Technology Madness
　................you will succeed!

PROMPT 13

A MANGLED MESS

How did my cell phone ear buds
become such a mangled mess?
I simply took them off my head
and put them on the desk!

When I went to put them on again
I became incredibly stressed!
For first I had to wrangle with them
putting my patience to the ultimate test!

And while I untangled this mangled mess
I imagined that while I was asleep,
They danced a wild Jitterbug
then fell into this jangled heep!

But I'm afraid that I will never know
the answer to this query,
For somehow it occurred to me,
It's like solving an Einstein Theory!

PROMPT 16

A Tech-Support Prayer

I said a little prayer today
before I picked up the phone,

"Please give me the strength
to speak to them
in a quiet and respectful tone!

Please let my rage be kept at bay
no matter how many hours
it takes today!

For in the end, it's all up to me
to respond to them,
pos-i-tive-ly!"

P.S. But that said, I know that it's a hard thing to do,
when the rage begins to boil inside of you!
Because you've been on the phone way too long,
and they still haven't a clue what the 'BLIP' is wrong!

Computer Madness

trapped within a little box
   relentless as a stubborn ox!
      slyer than the slyest fox!
         my words won't come out
            of my grey 'dell' box!

i push the print key
  just as i should,
    and wouldn't you think
      that if i would
          ....and did!

that the words
  should
    would
      print
if only they could!

but they can't
and don't
and won't
      in fact..........

PROMPT 19

SEATTLE GATES

my words are locked up
inside those seattle gates
  in that box that bill built
  that hold my words' fate!

and not even the techie
  could get past the ox
     or even the fox
     to open the box
    to print out my words

    how ab-surrrd!
              wouldn't you say?

        ....and so,

at the end of the day.......

who needs this keyboard
    in front of me!
from this computer madness
     i'll set myself free!

  with paper and pencil
    i'll write it all down!
    oh, my dear
      how simply profound!

to think that a style
  so old fashioned could be
    a far better way

    to communicate.................for me!

Google-Be-Goop

Google is great!
Google is good!
But does Google do
What it really should?

We type in Google
And Keyword a word,
And what comes up
Is excessive and absurd!

The first two choices
Might exactly be,
Exactly what you
Actually wanted to see!

But what about the rest
Of this extremely long list?
That causes you to shudder
Though you must persist!

The choices you have
Are 500 thousand and 53,
And to journey that search
Borders on In-san-i-ty!

But you start your search
Like a Sherlock Holmes' Case,
Opening one clue at a time
With such patience and grace!

Then suddenly you realize
You've spent an hour or more!
And you've already forgotten
What you were looking for!

So you return to Keyword
To type in more words
When you think to yourself
...............this is for the birds!

Oh, take me back to.............
Library Reference Desk days,
When life was less stressful
Before the Technology Craze!

PROMPT 22

Telephone Prompts

We pick up the phone
To make a simple call,
And what should be simple
Is not simple at all!

We just want to talk
To a living voice,
But sadly We don't seem
To have that choice!

For a machine suddenly
Begins to speak
And our blood pressure
Rapidly begins to peak!

For we know the routine
After too many years
Of the Telephone Prompts
Which drive us to tears!

Press one for English
Is the first thing we're asked,
And pushing that button
Is a most difficult task!

For how could it be?
In the U. S. of A.,
That speaking English
Is no longer the 'American Way'!

Press one for English
And for Spanish it's two
And now three for French
So tell us..............what's new?

It's not that we mind
The World stopping by,
But honestly!
Shouldn't they give English a try!

Once past the language
The prompting gets worse,
And we feel like we're stuck
In a 'Twilight Zone' curse!

But we must pay attention
To what's about to be said,
Or we'll have to start over
With anguished dread!

And sometimes an even
Worse fate then that!
..........if we don't make a choice
They'll disconnect us.....in fact!

And..............how RUDE is that?

PROMPT 24

No Passport Needed

I went on an Overseas trip today
But it's not quite what you think,
For I didn't pack any suitcases,
Or have to take the kitchen sink!

No need to call a travel agent
To purchase an airline ticket!
Or worry about a Passport,
Which can be a Sticky Wicket!

No notarized Birth Certificates
Were on my pre-travel list,
No luggage rules that drive me mad,
That painfully now exist!

The climate could be cold and snowy
Or perhaps be warm and sunny!
No trip to Thomas Cook was needed,
For the exchange of foreign money!

Now, imagine a trip without the TSA
Treating you like you've committed a crime!
Or the welcome relief of not having to snake
Through a wearisome mile-long line!

PROMPT 25

Lastly, as I prepare for this trip
I thought to myself with a sigh!
I don't have to learn another language,
As I fly through, THIS friendly sky!

For I know that English is what they speak
In all of these destinations!
But that's not to say I won't be without,
A variety of 'dialect' frustrations!

And one thing for certain would be the same
As taking an Airplane trip,
Was that by the end of this turbulent journey,
I'd be locked in a white-knuckle grip!

My patience would quickly be shredded
Into a trillion tiny pieces!
As my blood pressure goes bezerk,
And the panic within me increases!

If you still don't know where I'm headed
You may be the Luckiest Duck of them all!
For it may just mean in this Techno-World,
You've never made a Tech-Support call!

You're never quite sure where you're going
But you'll definitely know when you arrive!
Cause you'll be in some Foreign Country,
And the question is..............will you survive?

No Immunizations are needed
To protect you from catching a Bug!
You just need to have lots of patience,
For they're slower than an Escargot Slug!

They're polite, well-mannered and friendly
There's not a question in our mind!
The problem isn't in their behavior,
For their demeanor is always kind!

It's just that they seldom are able
To fix the problems that are at hand!
Though we tell them what those problems are,
They still don't understand!

So, if you thought for just a split second
That by air was the worst travel choice!
Just make another tech-support call,
And, on your next plane trip.........you'll Rejoice!

ESCARGOT CALL CENTER
OH NO!
T: Did you ask if I want garlic?
C: No, I asked if you can fix the problem...
T: I'm sorry, can you repeat that?

Poems – Then & Now

Back in the early 70s when I worked at the Pentagon, a building with approximately 26,000 employees (then), there were few places to go for privacy, hence the poem, Privacy! Thirty four years later, after a conversation with someone about cell phones and the fact that you can't even have privacy in restrooms anymore, I wrote, Privacy Revisited!

Privacy
(the Pentagon)

As I sit here
on the pot,
I find that I am
thinking a lot!

Not disturbed
by anyone......Oh!
This is the place
where I can go!
1972

Privacy Revisited
(Everywhere)

As I sit here
on the throne,
I find that I am
not alone!

For suddenly there's
that incessant tone,
Of the ever present
cellular phone!
2006

PROMPT 29

HEY DUDE! Can I go?
Hello? What movie do you want to see?
...
...
Okay, I'm looking at the movie section.
How rude...
Entertainment

Call The IT Police!

We've opened many more windows
for the robbers and the thieves,
And for all the sleezey No-Goodnicks,
with crime up their Spyware sleeves!

You'd think that surely our houses
and our cars would have been enough,
Without them also wanting to steal,
All our High-Technology STUFF!

But they crack and they hack
their way into our computers!
They're the Dreaded Tech Pirates
the Cyberspace Looters!

Like Robin Hood in Nottingham
they cannot be seen,
And without any warning
They'll steal your entire dream!

They'll keep sending you Spam
till you can't eat any more,
And then they'll send you Pop-Ups,
not sugar coated............for sure!

They go Phishing in places
where they shouldn't be,
And send us lots of Worms,
that we can't even see!

But there's someone out there
who can help keep the peace.
Can you guess who that is?
.........................it's the IT Police!

Quickly give them a call
so that you can get started,
Before all of your documents
join the Dearly Departed!

There are no gates or locks
or even keypad codes
For these IT Police
use a much different mode!

It's Pestblock and Adware
and Spinach and Bots!
And if you think that's all,
there's still lots and lots!

There's Sitehound for Firefox
to protect you from Scams!
This is all beginning to sound,
like Green Eggs and Ham!

And what about the Trojans
who come over to attack?
And the Browser Hijackers
and the Adbots..........what's that?

My head is now spinning
so I think I'll escape!
Before I catch a Virus,
or become Phishing Bait!

Good Luck in your attempts
while you're out on your search!
I hope you find something
And aren't left out in the Lurch!

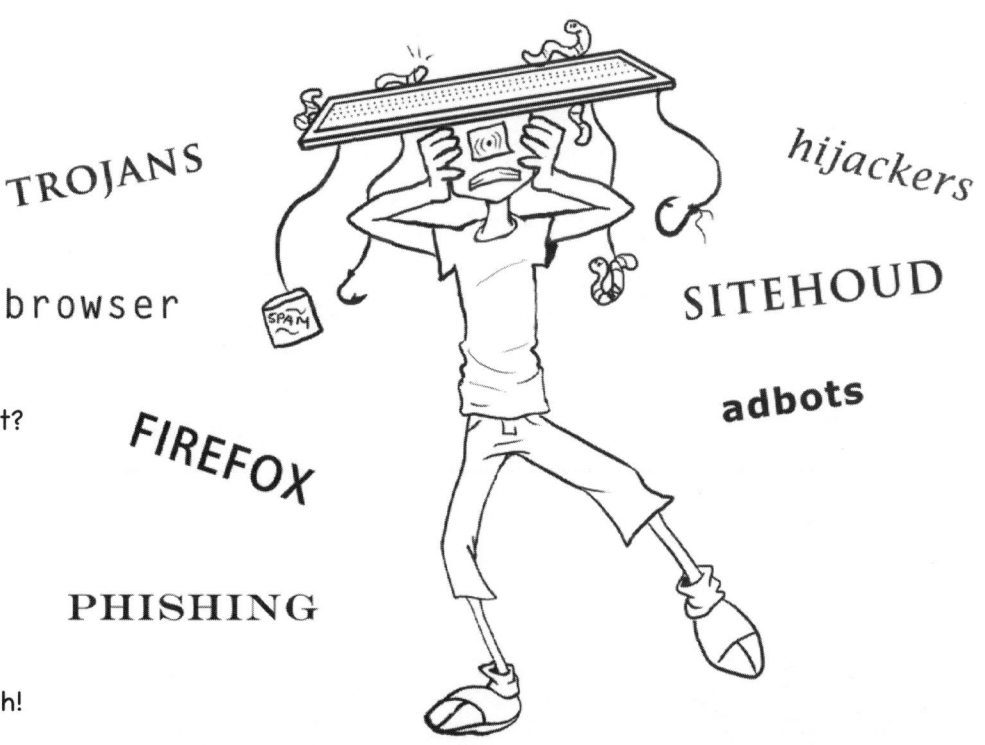

TROJANS

browser

FIREFOX

PHISHING

hijackers

SITEHOUD

adbots

PROMPT 32

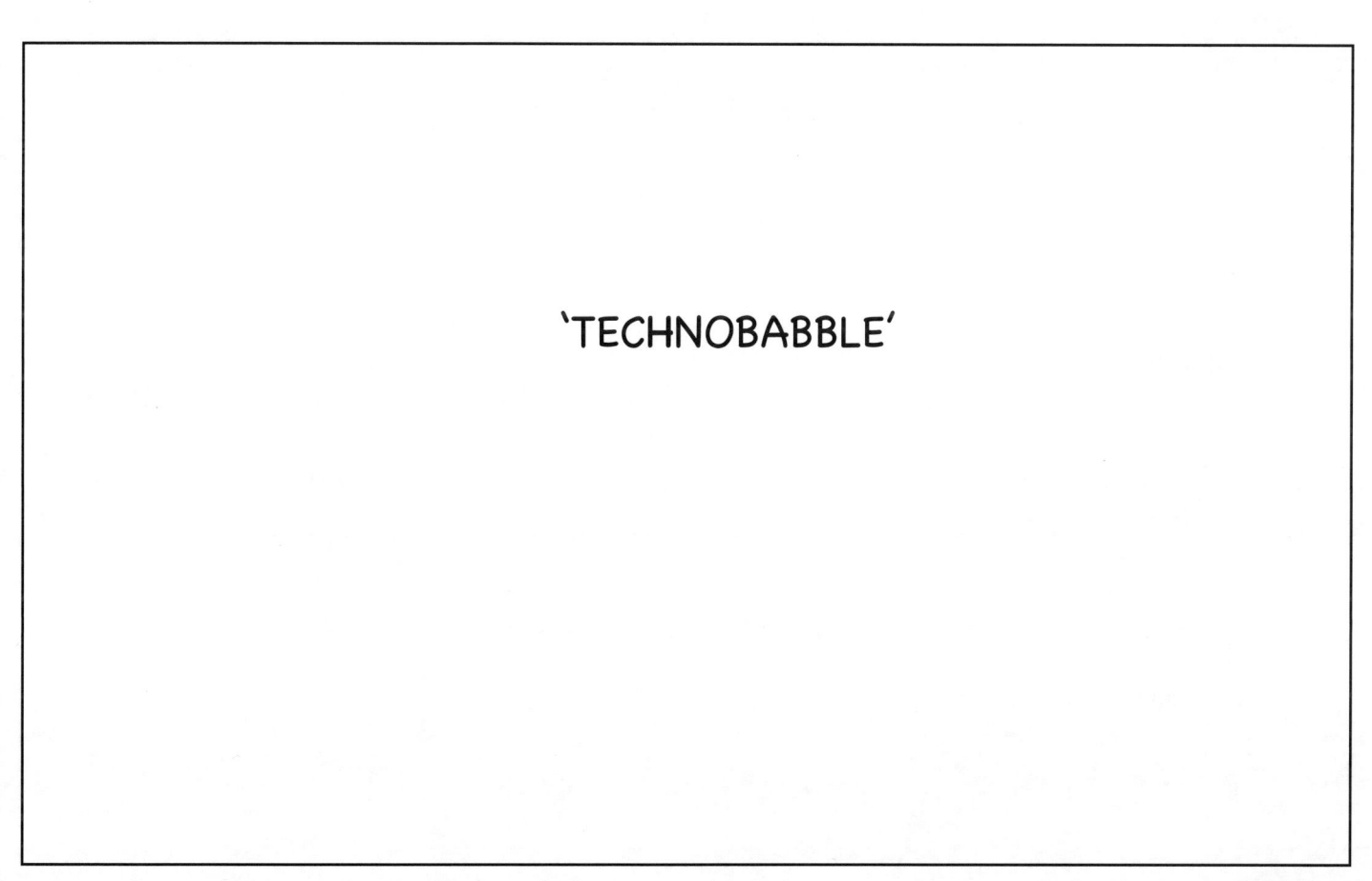

'TECHNOBABBLE'

'TECHNOBABBLE' DEFINITIONS

Dictionary
Technobabble - |ˈteknōbabəl| |tknouˈbɵbəl|
noun informal
incomprehensible technical jargon.

urbandictionary.com
Technobabble – The art of using a bunch of nonsensical ramblings to solve of explain a problem. Usually used by SCI-FI writers once they have written themselves in a corner.
Scotty couldn't beam Kirk up until he reversed the polarity on the microconvertor to align it in phase with the eisenberg compensator.

netlingo.com
Technobabble – A series of high-tech terms strung together to sound impressive without actually meaning anything. Can be used to overwhelm newbies and give a false sense of superiority to people in the industry who use such jargon.

ANSWERS TO 'TECHNOBABBLE' MATCH GAME

(on the following pages)

PROMPT			PROMPT	
1.	D		8.	H
2.	F		9.	I
3.	C		10.	J
4.	A		11.	K
5.	B		12.	L
6.	G		13.	M
7.	E		14.	N

'TECHNOBABBLE' MATCH GAME

1. zipperhead
2. murk
3. muggle
4. mainsleaze
5. jabber
6. byte-bonding
7. newbie

PROMPT 37

DEFINITIONS

A. A company who goes to the 'dark side' and spams their customers.

B. Transmit meaningless data via networks or keyboards when chatting.

C. Someone who is just plain mundane.

D. A person with a closed mind.

E. Someone who is new to the internet or to computers in general.

F. Disclaimer at the end of spam email assuring reader, message complies with a bill that was put before the Senate.

G. When computer users get together and discuss things that non-Computer users don't understand.

'TECHNOBABBLE' MATCH GAME

8. noob

9. wetware

10. stickiness

11. sticky menu

12. scrunch

13. stomp on

14. baud barf

DEFINITIONS

N. The strange noises one hears when a computer is connecting to a network.

M. To overwrite by accident.

J. How long the average user spends at a website.

l. To break through security to gain unauthorized access to a private network.

K. A menu that will stay open if the pointer is put in the correct position.

H. Jerks who spend their time flaming others because they think they are better than everyone else.

I. Another name for your brain.

JAPANESE 'TECHNO'

HAIKU POETRY
(ERROR MESSAGES)

JAPANESE SMILEY FACES

PROMPT 42

Haiku Poetry

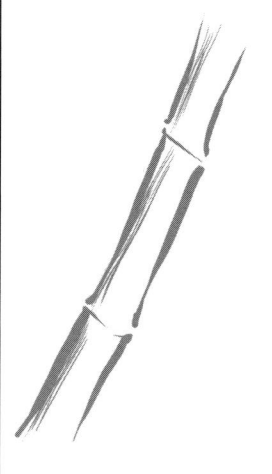

Definition of Haiku Poetry:

haiku |'hi,koo; hi'koo | | haI'ku | | hAku:|
noun (pl. same or -kus |hakuz|)
A Japanese poem of seventeen syllables, in three lines of five, seven, and five, traditionally evoking images of the natural world.
• An English imitation of this.

ORIGIN Japanese, contracted form of ***haikai no ku 'light verse.'***

NOTE: On the following pages you will find Japanese Haiku Poetry used to express 'computer error messages' that are often rude, obnoxious and abrupt, i.e., "This computer has committed a fatal error and will now shut down"………..BLEEP!

The first six pages of Haiku Poems were found on the Internet in, netlingo.com, under 'H' for Haiku (no kidding)! They also, 'popped up' on several other sites. It appears that they were originally written as submissions to a Haiku contest for, Salon 21 Magazine. I have added the authors names to the ones I found and apologize to the rest!

The next thirteen pages were written by me at 1:30 a.m. after Googling and finding a long 'error message' list (colba.net)! While googling and writing, I was giggling out loud! Hope they bring you a giggle or more!

Haiku Poetry

A file that Big?
It might be very useful.
But now it is gone.
  (David J. Liszewski)

The Web site you seek
Cannot be located, but
Countless more exist.
  (Joy Rothke)

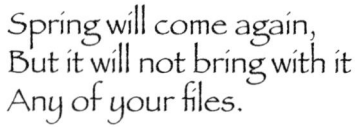

Spring will come again,
But it will not bring with it
Any of your files.
  (Cheryl Walker)

Chaos reigns within.
Reflect, repent, and reboot.
Order shall return.
  (Suzie Wagner)

PROMPT 45

Windows NT crashed.
I am the blue screen of death.
No one hears your screams.
    (Peter Rothman)

Yesterday it worked.
Today it is not working.
Windows is like that.
    (Margaret Segall)

First snow, then silence.
This thousand-dollar screen dies
So beautifully.
    (Simon Firth)

With searching comes loss
And the presence of absence:
"My Novel" not found.
    (Howard Korder)

PROMPT 46

Haiku Poetry

The tao that is seen
Is not the true tao – until
You bring fresh toner.
    (Bill Torcaso)

Stay the patient course.
Of little worth is your ire.
The network is down.
    (David Ansel)

A crash reduces
Your expensive computer
To a simple stone.
    (James Lopez)

Three things are certain:
Death, taxes and lost data.
Guess which has occurred.
    (David Dixon)

PROMPT 47

You step in the stream,
But the water has moved on.
This page is not here.
    (Cass Whittington)

    Rather than a beep,
    or a rude error message,
    these words: "File not found."
        (Len Dvorkin)

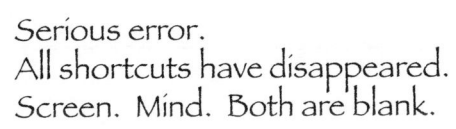

        Serious error.
        All shortcuts have disappeared.
        Screen. Mind. Both are blank.
            (Ian Hughes)

  Out of memory.
  We wish to hold the whole sky,
  But we never will.
      (Francis Heaney)

PROMPT 48

Haiku Poetry

Login incorrect.
Only perfect spellers
may enter this system.
   (Jason Axley)

The ten thousand things
How long do any persist?
Netscape, too, has gone.
   (Unknown)

Printer not ready
Could be a fatal error
Have your pen handy?
   (Pat Davis)

I ate your Web page.
Forgive me. It was juicy
And tart on my tongue.
   (Unknown)

PROMPT 49

Seeing my great fault
Through darkening blue windows
I begin again.
  (Chris Walsh)

Having been erased
The document youre seeking
Must now be retyped.
  (Judy Birmingham)

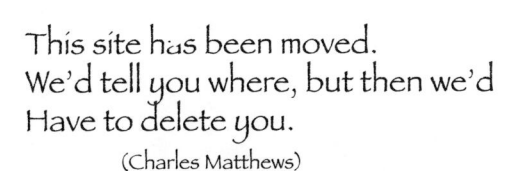

To have no errors
Would be life without meaning
No struggle, no joy.
  (Brian M. Porter)

This site has been moved.
We'd tell you where, but then we'd
Have to delete you.
  (Charles Matthews)

PROMPT 50

Haiku Poetry
by
patricia stutts

The Haiku Poems on the following pages are on the subject of 'error messages'. Each Haiku Poem will have a set of quotation marks around the line that is an authentic "error message!" There is one exception..........can you figure out which one is not an exact quoted 'error message'? Answer is found at, www.patafeabooks.com!

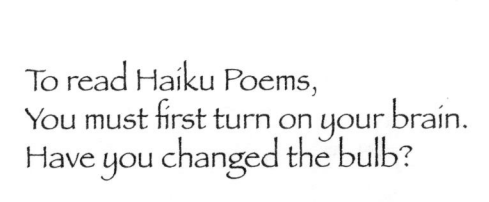

To read Haiku Poems,
You must first turn on your brain.
Have you changed the bulb?

PROMPT 52

Haiku Poetry

"Hold down start button."
Entering frozen Tundra.
See you back shortly.

        Your choice is not bright.
        "Rainbow pinwheel goes bonkers."
        Your gold is now lost.

                "Connection not found."
                No responsibility.
                Ostriches by birth.

    "User is unknown."
    Who is this Mailer-Daemon?
    Is he unknown, too?

PROMPT 53

You read directions.
"Your entry was invalid."
Who taught them the rules?

What are you thinking?
"An unknown error occurred."
Try one that you know!

"Boot has failed in drive."
Next time wear regular shoes,
Try it in third gear!

Marathon needed!
"Device needs to run Windows."
What about Boston?

PROMPT 54

Haiku Poetry

Excess multi-task!
"Too many files opened."
It was a nice try!

"Device not ready."
Did you make an appointment?
Next time bring a snack!

Is your brain failing?
"There's not enough memory!"
Need to upgrade both!

Are we talking cars?
"Error loading Explorer."
I have a Chevy!

PROMPT 55

What did I do wrong?
"Fatal error......shutting down!"
Please don't die on me!

I have been prepared.
"Fatal error.........shutting down!"
The grave has been dug!

Drive C speaks to you.
"You have run out of disk space!"
Then go to the store!

Can't open windows.
"Insufficient memory."
Home.Office Depot!

PROMPT 56

Haiku Poetry

"Object reference,
Is null or not an object!"
A genius wrote this!

"Packed file corrupt!"
Has it been out of your sight?
TSA inquires!

"Permission denied!"
Well, who asked you anyway!
Mind your own business!"

Try a different name.
"The folder does not exist!"
I created it!

PROMPT 57

Outlook Express says,
"The message could not be sent."
Try the Post Office!

Changes every day.
"The network path was not found."
Then try the highway!

Tell USB Port,
"The printer cannot be found!"
Maybe scuzzy knows?

Think ahead next time.
"This Web page could not be saved!"
Hire a lifeguard!

Haiku Poetry

"To use Net Passport
You must enable cookies!"
I'm on a diet!

Take another tact.
"Logical Assertion Failed!"
Try illogical!

Do the next best thing.
"Out of environment space!"
Call Al Gore for more!

Try other options.
"Out of Resources Error."
Try out-sourcing it!

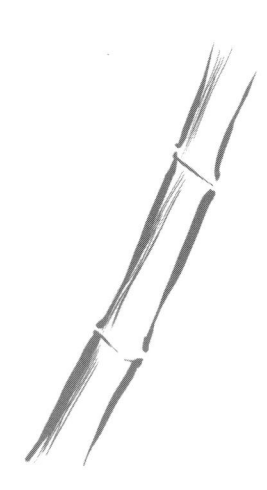

What are your hours?
"Server unavailable."
Use a Post-It Note.

"Signal out of range."
There are no towers allowed!
Ask the Buffalo!

Identify it.
"The object could not be found!"
Try my affection!

"Request has time out."
Why did you set the timer?
Another chance......please!

PROMPT 60

Haiku Poetry

Must be X-Rated!
"The page cannot be displayed!"
Reopen as tiff!

Who you gonna call?
"Word could not fire event."
Then get Donald Trump!"

Where did it get lost?
"This file cannot be found."
Inquiring minds.

You must have all day!
"Check the path and try again."
I am out of time!

PROMPT 61

Before you begin,
"Do you want to run setup?"
I prefer to walk!

Keep trying your task.
"Cannot create the item!"
You have six days left!

Where is the party?
"Unable to locate host."
Not necessary!

Why are you worried?
"Unable to locate host"
Just need food and drinks!

PROMPT 62

Haiku Poetry

Papers were filed.
"This file contains no data."
Expectations high!

Password required.
"Connection refused by Host!"
Will a pink slip work?

Wanted, a Big Mac!
"Resources are limited."
I'll take a mini!

"No such dialect!"
Is this a tech-support call?
I don't understand!

PROMPT 63

It's my property!
"Access not allowed!"
Your name isn't Hal?

Who did you talk to?
"Expected a reference!"
Go to Library!

I pushed send button.
"Reply has not yet arrived!"
One second has passed!

Need not look further.
"Some data was the wrong type."
Am I the right type?

PROMPT 64

Haiku Poetry

How do you know this?
"This file was not completed."
Do you have a list?

Do you need glasses?
"Some data could not be read!"
Is it in English?

"Enter a password!"
I need some privacy please!
Promise you won't peek!

"Installation failed!"
Was it given a fair test?
Take the class again!

It is 3:00 a.m.
"My eyes lids are shutting down!"
Time to press Sleep Mode!

Thank you for reading.
"Haiku error messages."
Hope you enjoyed them!

PROMPT 66

JAPANESE 'SMILEYS'

Japanese Smileys

A derivative of the emoticon, (American Smileys, which just celebrated their 25th Anniversary in May of this year), this form of Smiley is a sequence of typed characters that creates a facial expression. As is the case with straight-on-smileys, Japanese Smileys do not require users to tilt their heads sideways in order to see them. These Smileys are developed by Internet users in Japan and even though they reflect cultural nuances, they are fast becoming common forms of online expression worldwide. They were thought to have first appeared in May of 1985, created by a nuclear scientist, (~_~). From that time, they have continued to grow! (netlingo)

(^_^) male smiley   (^.^) female smiley   (^L^) or (^(^) happy

(-_-) secret smile   (^o^) laughing out loud

(^_^;) laughing to cover nervousness   (^_^)/ waving hello

(;_;)/ waving goodbye   (v_v) expressionless

PROMPT 69

(^_~) or (^_-) winking   (*^o^*) or (*^.^*) exciting

\(^_^)/ joyful   (;_;) or (~~>.<~~) crying   (>.<) or (>_<) angry

(^o^;> excuse me?   (*^_^*) blushing (or shy)   (@_@) stunned

(^_^;;;) embarrassed (or in a cold sweat)   (?_?) confused (or wondering)

(!_!) or (o_o) shocked   (*_*) frightened (or in love)

(=_=)~ sleepy   (u_u) sleeping

'\=o-o=/' wearing glasses   m(_)m humble bow of thanks or apology

Sayonara (-_-)/

PROMPT 70

SMILEYS AND EMOTICONS

:-(:-)

THE BIRTH OF THE SMILEY

HAPPY 25TH BIRTHDAY!

The official birth date of the Smiley is September 19, 1982. It was created by, Scott E. Fahlman, at Carnegie Mellon University. The two original "glyphs" by Scott were, :-) and :-(Since Scott posted his first Smiley proposal, many other Smileys have been devised by many others. (Go to www.cs.cmu.edu/~sef/sefSmiley.htm, to read an Article by Scott E. Fahlman.)

A Smiley is a sequence of characters on your computer keyboard. If you don't see it, try tilting your head to the left -- the colon represents the eyes, the dash represents the nose and the right parenthesis represents the mouth. Smileys usually follow after the punctuation (or in place of the punctuation) at the end of a sentence. A Smiley tells someone what you really mean when you make an offhand remark. They are also called, 'Emoticons', because they intend to convey emotion! A new generation of Smileys has appeared on the scene and netLingo is fast trying to track them down. They consider these, 'straight-on smileys', as another form of ASCII ART: those in which you do not tilt your head but rather look at it straight on. (If you want to find more fun categories of Smileys, a little risqué, go to netlingo.com and look for information on Smileys.......there is a link to them.

PROMPT 73

STRAIGHT-ON SMILEYS

Butterfly	}|{	Cat	=^..^=
Fish	<><	Big Hug	(((H)))
Koala	@(*0*)@	Dazzling Grin	*^_^*

SIDEWAYS SMILEYS

Christmas Tree	*<<<<+	Rudolph	3:*>
Computered Out	=%-O	Bored	:-!
Up All Night	%-|	A Rose	@>--;--
Angel Wink/Female	O*-)	Angel Wink/Male	0;-)
Jim Carrey	?:^[]	Bill Clinton	=|:o}

PROMPT 74

'TECHNO'

LIMERICKS

by
patricia stutts

DEFINITION OF A LIMERICK

Limerick |ˈlim(e)rik| | lIm(e)rIk | | lIm(e)rIk|

Noun
A humorous, frequently bawdy, verse of three long and two short lines rhyming, aabba, popularized by Edward Lear. ORIGIN late 19th cent.: said to be from the chorus "Will you come up to Limerick?", sung between improvised verses at a gathering.

PROMPT 77

Version I:

I have written these Limericks for thee!
Defining words from Technology.
   They can't be naughty!
   Or even bawdy!
For children might read them......dear me!

Version II:

I have written these Limericks you see!
Defining words of 'techno'...........lo-gy!
   Like who ever knew,
   That I could accrue,
Enough knowledge for a Technology Degree!

PROMPT 78

PROMPT 79

TECHNOSTRESS

In a world of increased TECHNOSTRESS!
From Technology not working.......I guess!
    We're all in a tizzy!
    And definitely dizzy!
From the Technology on which we obsess!

            TECHNOSIS

       There once was a condition called TECHNOSIS!
       For which there was just one prognosis!
           That our dependency for,
           The Technology lure,
       Could only be cured with Hypnosis!

TECHNOPHOBIA

TECHNOPHOBIA, oh what could that be?
It afflicts many people..........even me!
   Throughout the World over,
    From Hong Kong to Dover,
It's the fear of Technology............you see!

Mainsleaze

Murk

Jabber   Newbie

ZIPPERHEAD

TECHNOBABBLE

TECHNOBABBLE is the language of the techies!
Even spoken by Spock and Star Trekies!
   The jargon's far out!
    Don't know what it's about?
And wonder if it's used by the Shrekies?

PROMPT 82

TECHNOPHILIC CLAN REUNION

TECHNOPHILIC

There once was a TECHNOPHILIC man!
Who was part of a much larger Clan!
   And with every new gadget,
     Which to them was sheer magic,
To own it was part of their plan!

TECHNOHEDONIST

The TECHNOHEDONIST has got to promote!
So he can stand there and proudly gloat!
    About the next best thing!
    That Technology will bring!
Even if the product won't float!

Sniff Sniff YUM! Peking Duck!

NOTE: Product in lake that didn't float was, the iSmell, which was supposed to plug into your computer with a USB Port and allow you to smell scents when surfing the Web, i.e., in a flower Web site, you could smell roses, perfume Web site, Channel, perhaps, Chinese Take-Out, Peking Duck!

PROMPT 84

TECHNOCOWBOYS

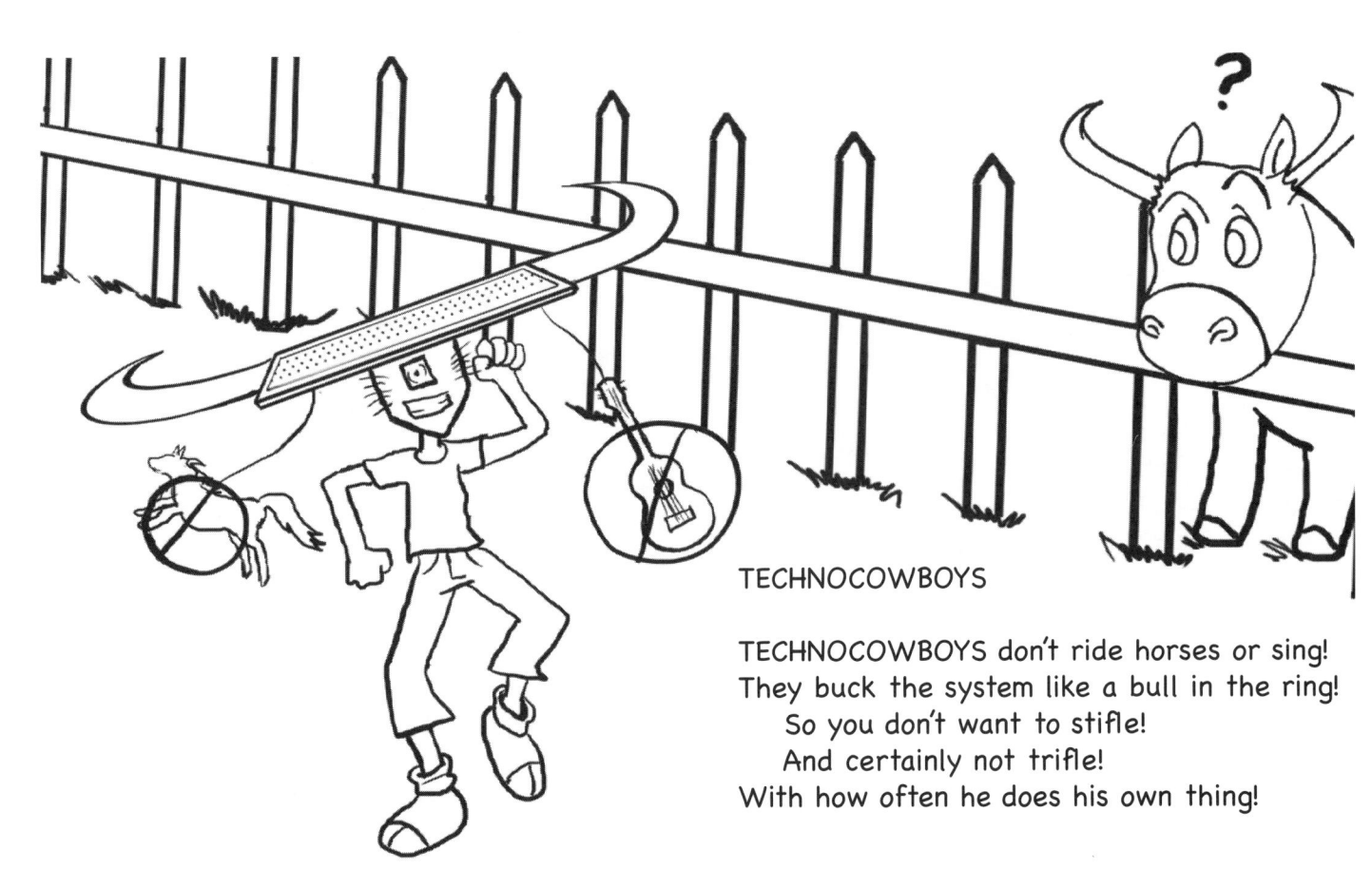

TECHNOCOWBOYS don't ride horses or sing!
They buck the system like a bull in the ring!
   So you don't want to stifle!
   And certainly not trifle!
With how often he does his own thing!

TECHNOHIPPIES

TECHNOHIPPIES don't wear bell bottoms or beads!
And probably don't even smoke weeds!
   In cyberspace they're cool!
   They're the hackers who rule!
And tell the older generation to concede!

PROMPT 86

TECHNOJUNKIE $

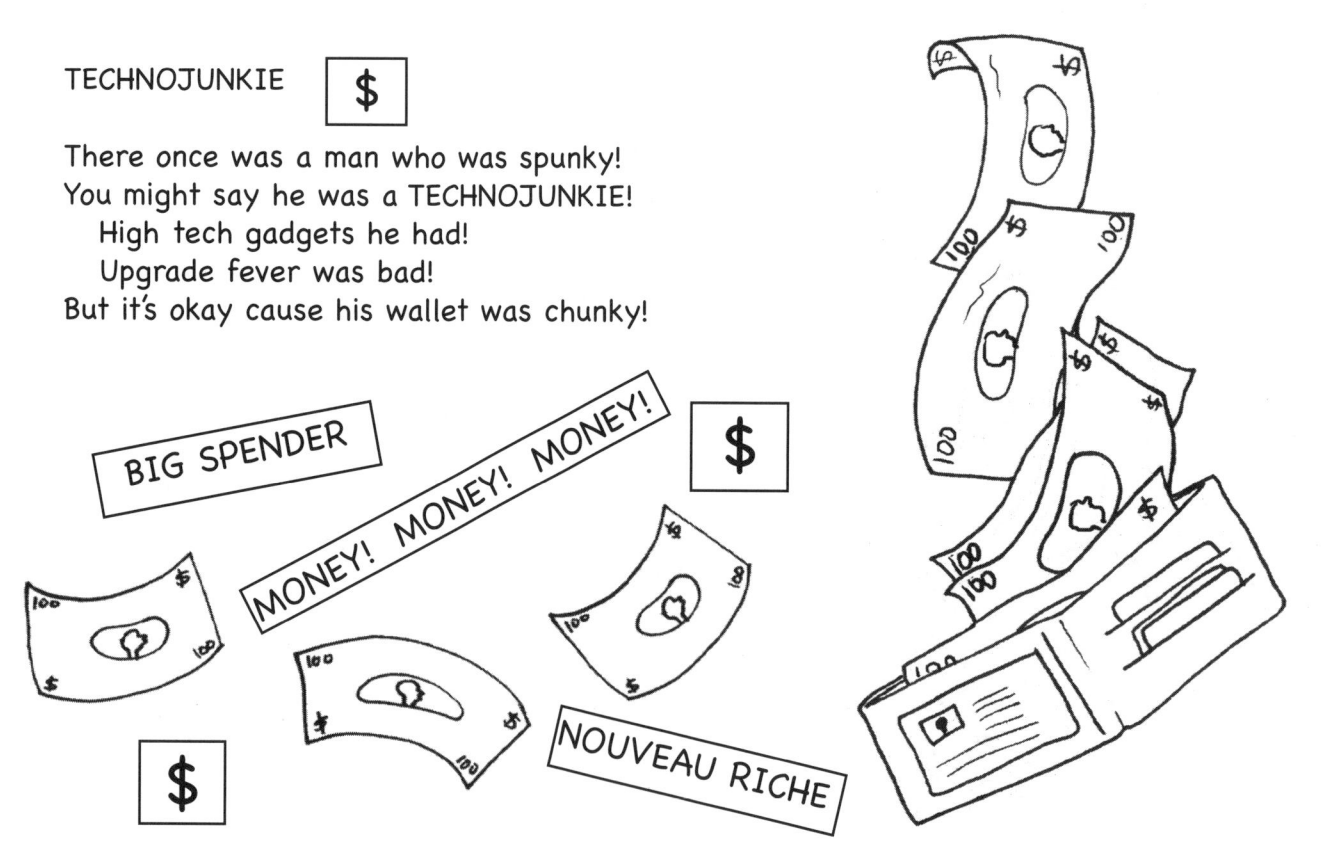

There once was a man who was spunky!
You might say he was a TECHNOJUNKIE!
    High tech gadgets he had!
      Upgrade fever was bad!
But it's okay cause his wallet was chunky!

BIG SPENDER

MONEY! MONEY! MONEY!

$

$

NOUVEAU RICHE

DREAD!

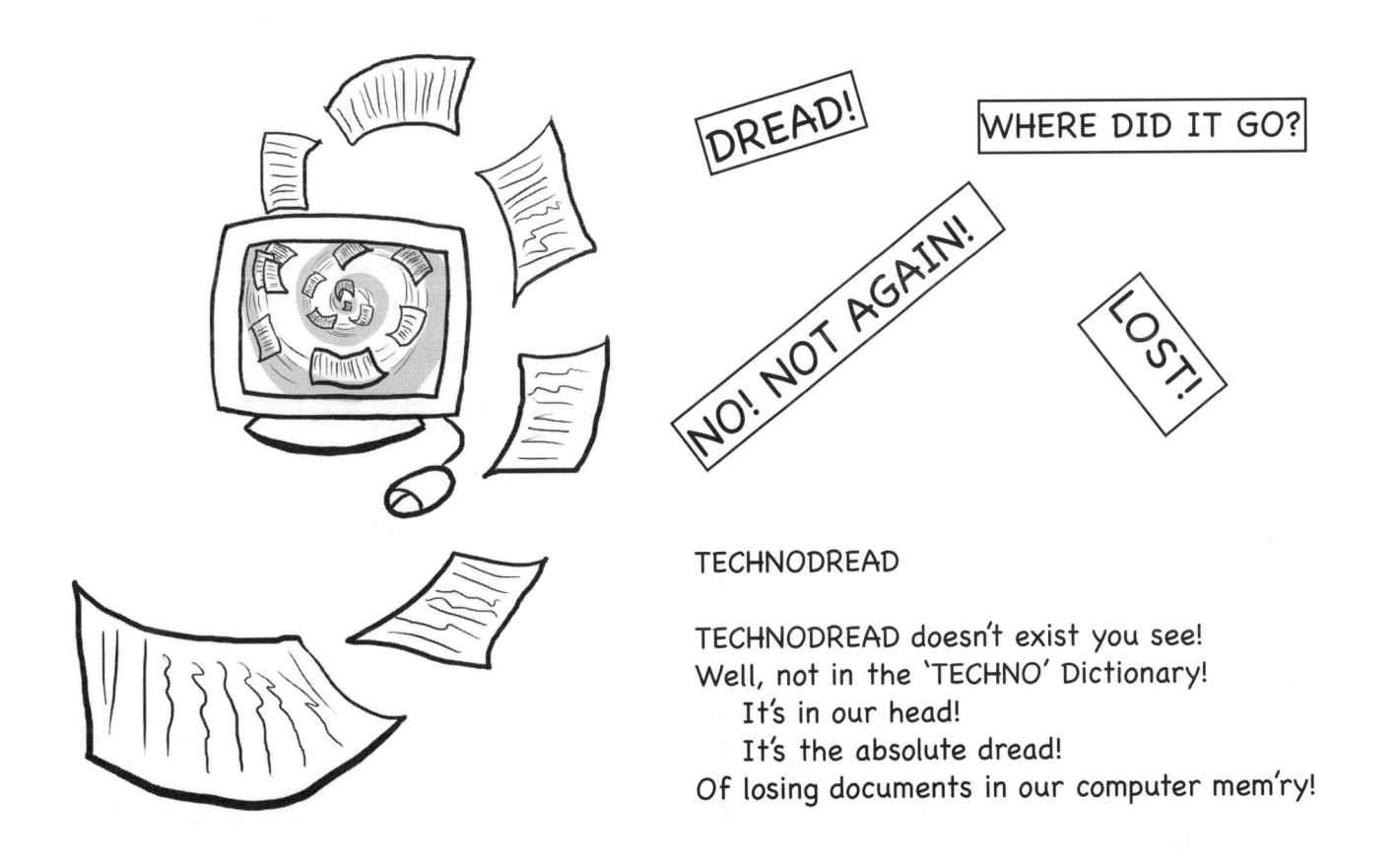

WHERE DID IT GO?

NO! NOT AGAIN!

LOST!

TECHNODREAD

TECHNODREAD doesn't exist you see!
Well, not in the 'TECHNO' Dictionary!
    It's in our head!
    It's the absolute dread!
Of losing documents in our computer mem'ry!

PROMPT 88

TECHNOHYSTERIA

A man with TECHNOHYSTERIA!
Was like someone with a case of Maleria!
    He was out of control!
    A once bitten soul!
In a Technology Laddened Area!

PROMPT 89

TECHNOCREEP

When your boundaries no longer have a line!
And trivial gadgets devour your time!
    Listen for the beep!
    Of TECHNOCREEP!
Before it consumes your life and your mind!

TECHNOSKEPTICAL

A TECHNOSKEPTICAL was in a Mega Store!
With cameras, computers and accessories galore!
    So he asked himself?
    If what's on the shelf?
Were things he needed or could actually ignore!

TECHNOBOLLOX

Inaccurate drivel one might accurately say!
Is TECHNOBOLLOX in a most direct way!
    It's frenzied and spasmodic!
    With buzzwords that are exotic!
That sound like hogwash at the end of the day!

PROMPT 92

TECHNOFLID

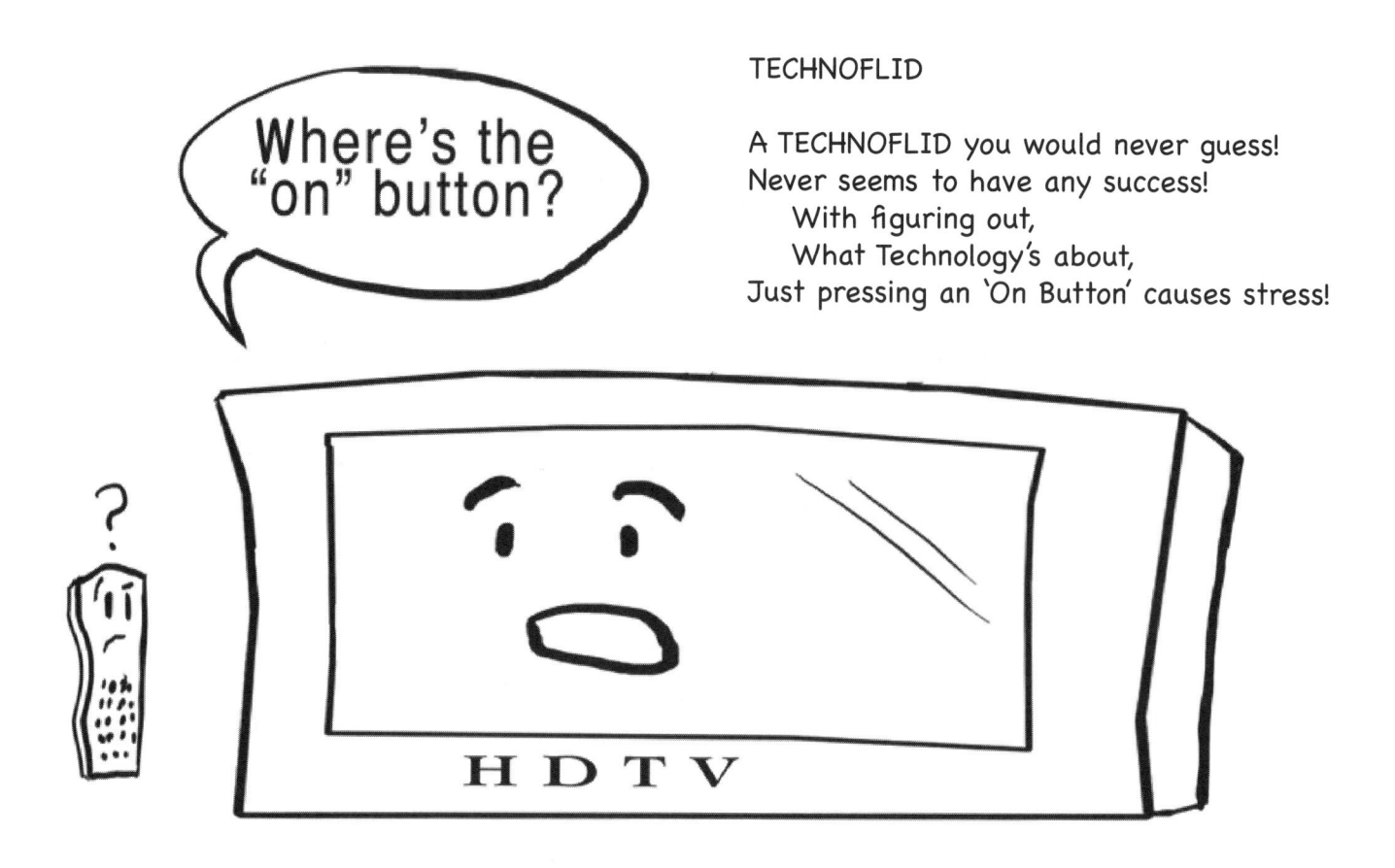

A TECHNOFLID you would never guess!
Never seems to have any success!
    With figuring out,
      What Technology's about,
Just pressing an 'On Button' causes stress!

"zing went the bling ♥ of my heart"

TECHNOBLING

TECHNOBLING is not just any old thing!
Not even a Diamond or Sapphire Ring!
    It's anything magic!
      In the form of a gadget!
That's expensive and has Zest and Zing!

PROMPT 94

TECHNOHOLIC

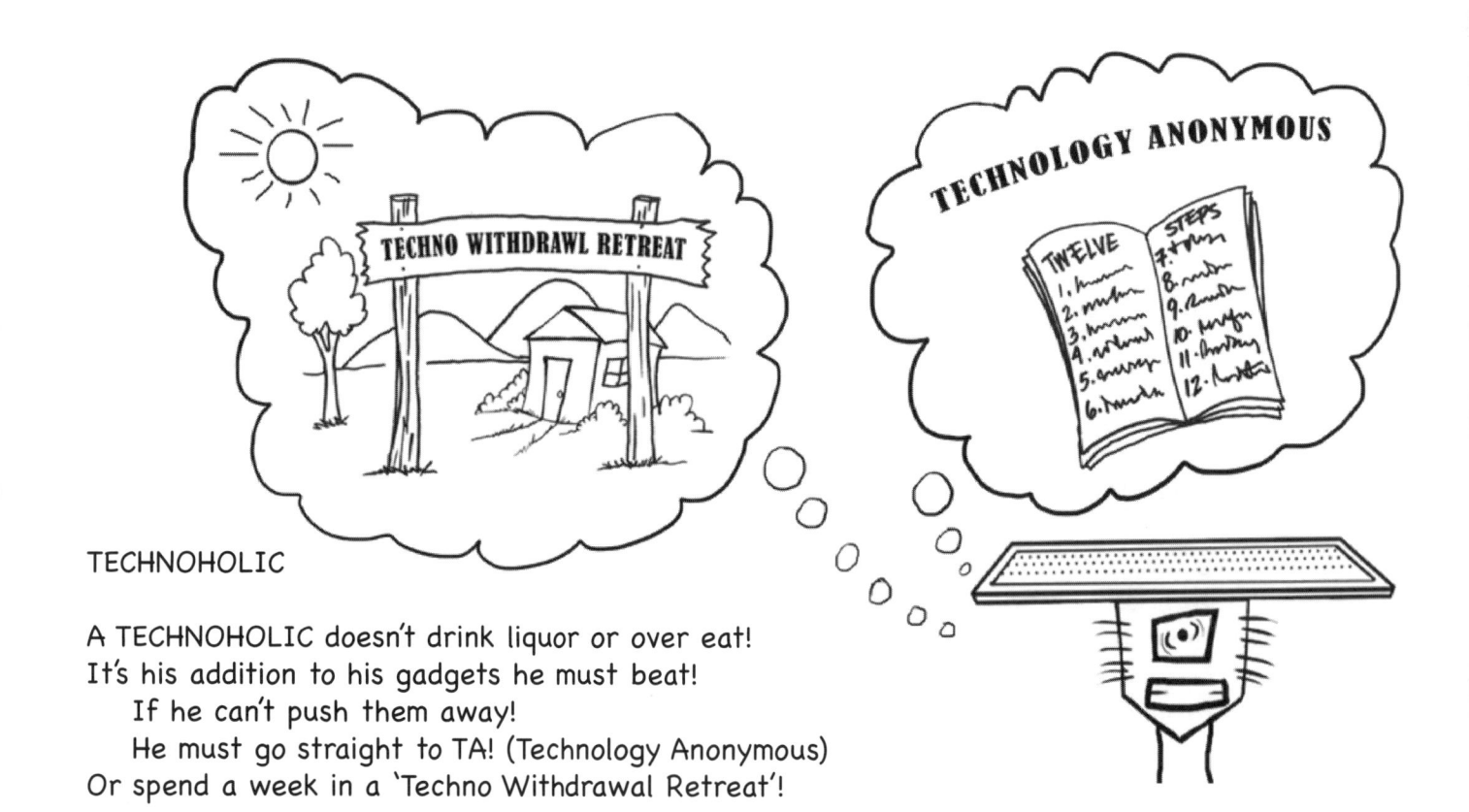

A TECHNOHOLIC doesn't drink liquor or over eat!
It's his addition to his gadgets he must beat!
    If he can't push them away!
    He must go straight to TA! (Technology Anonymous)
Or spend a week in a 'Techno Withdrawal Retreat'!

PROMPT 95

'ELECTRONIC PICKET'

STOP! I CAN'T TAKE ANYMORE! I GIVE UP!

YOU'VE GOT MAIL

TECHNOSTRIKE

The TECHNOSTRIKE was a union labour action.*
That gave its supporters such grand satisfaction!
    It wasn't a game of cricket!
    It was an 'electronic picket'!
Jamming emails with their organized faction!

*This term was used for a Strike at a company in Wales, Great Britain. (UrbanDictionary)

TECHNOMUSIC

TECHNOMUSIC started around the year 1970!
With Kraftwerk, from Düsseldorf, Germany!
    Juan Atkins, they say,
        Who was a Detroit, DJ!
Developed U.S. TECHNO in the year 1980!

PROMPT 97

TECHNOMUSIC has an electronic synthesizer base!
You can even 'Techno Sweat' to its 160 bpm pace!
    It came after Rock 'N Roll!
    Draws from Funk and from Soul!
And to dance to it................you need lots of space!

(Go to, kraftwerk.com to listen to their sound and Google Juan Atkins and techno music to hear and learn more!)

PROMPT 98

TECHNOHUMORIST

A TECHNOHUMORIST I am................am I?
Well, at very least I'm giving it a try!
    With the madness of it all!
      I'm trying to have a ball!
But sometimes I sit down and just cry!

PROMPT 99

TECHNOME

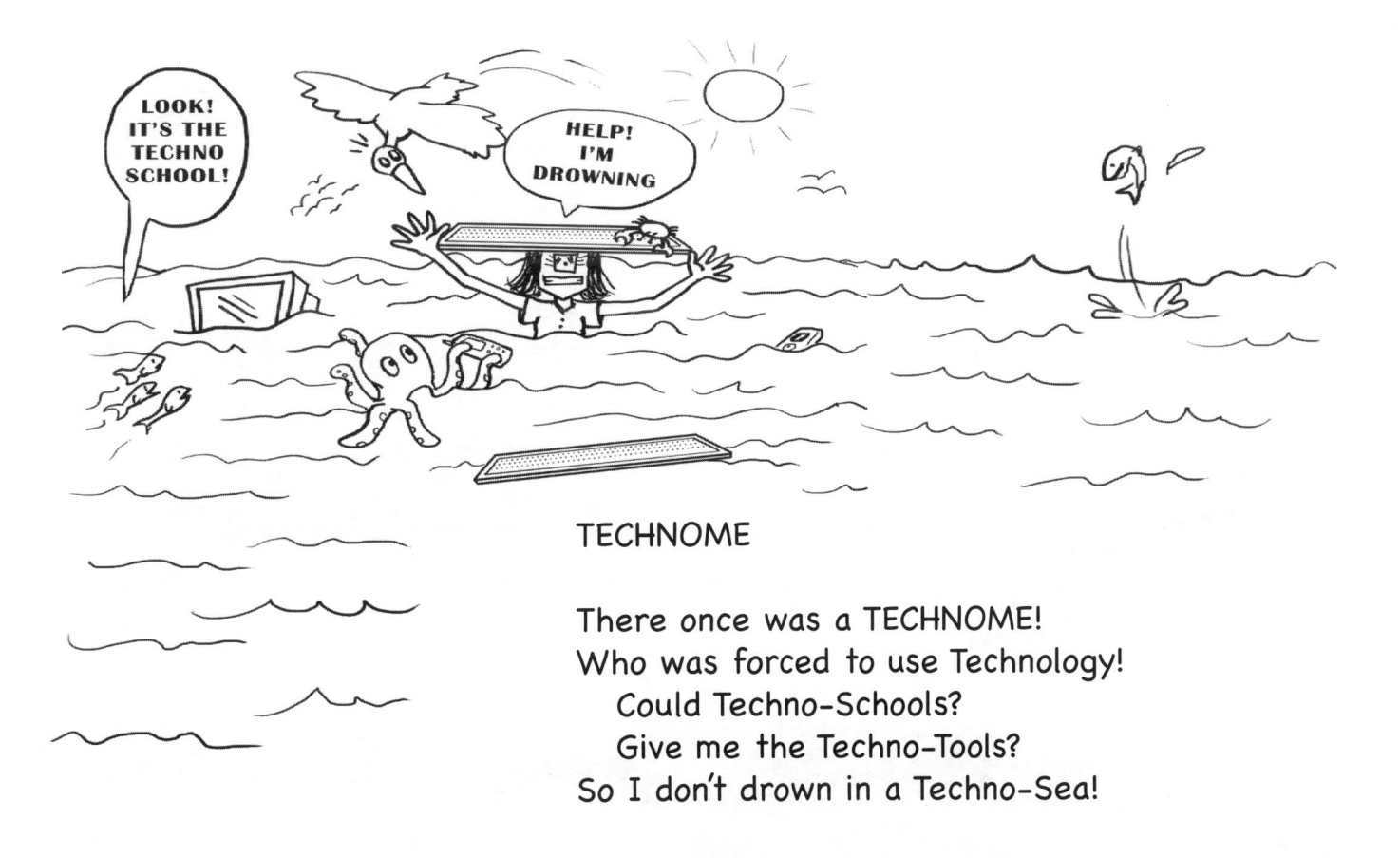

There once was a TECHNOME!
Who was forced to use Technology!
   Could Techno-Schools?
   Give me the Techno-Tools?
So I don't drown in a Techno-Sea!

PROMPT 100

TECHNOJOCK

Have you ever met a TECHNOJOCK?
They think they're Cool cause they know a lot!
    In truth, they do!
      And, if allowed to!
They'll strut around like an Arrogant Peacock!

THE TECHNOJOCK STRUT

PROMPT 101

MORE 'TECHNO' WORDS
NOT WRITTEN FOR!

TECHNOSTALGIA - Nostalgia for the simpler forms of Technology!

TECHNOREALIST - Someone who has a balanced and realistic view of Technology!

TECHNOCRAT - A person who is part of the technically skilled elite!

TECHNOGYPSY - Someone who travels with all their gadgets!

TECHNOCHONDRIA - Always thinking there is something wrong with their gadgets!

TECHNOVATOR - Boring Techno Music - cross between Techno and Elevator Music!

TECHNOTARD - Someone who couldn't operate a PC with a Quick-Start Manual!

TECHNOATHEIST - A person who realizes that there is no supreme form of Technology, they all have their challenges.

BLOG

BLOGS
BLOGGED
BLOGGER
BLOGGLE
BLOGGLED
BLOGSPOT
BLOGROLL
BLOGGING
BLOGOSPHERE
BLOGSPACE
BLOGISTAN
BLOGLAND

BLOG

DEFINITION OF A BLOG:

blog |bläg| |blag|
noun a Web site on which an individual or group of users produces an ongoing narrative : *Most of his work colleagues were unaware of his blog until recently.*
verb (**blogged, blogging**) [intrans.]
add new material to or regularly update a blog.

DERIVATIVES
blogger noun
ORIGIN a shortening of **weblog** - ORIGIN 1990s: from web in the sense [World Wide Web] and log in the sense [regular record of incidents.]

FYI: More than 100 million Blogs are being tracked…………that just Bloggles my mind! Way too much information!

PROMPT 105

THE BLOGGERS' GOG*

NOTE: Keep in mind that the Bloggers' Gogs that I have chosen to use in this Section, do not necessarily reflect my personal opinion. They are just the Blogs that I thought evoked a good laugh or chuckle, as they reflect on the reality of what is going on in the world of Technology!

Blogs are an interesting 'byte' of history that someday we will look back on............but, I'm not sure if we will look back on them and still laugh because we truly thought that it was humorous or if we will look back on them and cry with tears of joy(ful) memories at the humor in the Madness of Technology!

For the record, I am not singling out Apple! I think they are a great Company with the most innovative product design in the Industry!

*GOG – What blogs out of a Blogger's mouth!

PROMPT 106

THE BLOGGERS' GOG

In September 2007, just two months after the release of the, iPhone, Apple announced a $200 price reduction, taking the price from $599 to $399. A flood of complaints from the Technohedonists (wow, get to use one of my new 'Techno' terms), prompted Apple to offer early purchasers of the iPhone, a $100 in-store credit!

Early iPhone users and non-Apple addicts vented their emotions (good, bad and indifferent) about the price cut. (I make no comment on the grammer, except this one, of course! Whatever happened to, Spell Check!)

These Bloggers were responding to the above announcement:

- i agree that they should get over it. EVERYBODY, well the smart ones anyway know that you wait till expensive electonics drop the price cause they don't do it to long after it comes out. yea 2 months is short but sucks they should have waited. if i were th CEO i woodn't give them nothin back, loosers, yea who said they were spoiled brats i agree w/ that too. AMEN

- My math is old fashion, but $599 minus $399 equals $200 not $100!

- He who waits gets the good deal... He who doesn't gets screwed.

PROMPT 107

- I would like to thank Apple for coming out with their over-hyped piece of icrap phone. The price cuts and rebates that resulted made purchasing my Blackberry Curve an extremely good buy. Thanks Apple!!

- Just like most of us Americans..spolied and upset if they don't get their way.

- Steve says... all apple buyers takeoff your clothes and put your underwear on your head Steve says ... buy an i Phone and we will turn it into a brick if you hack it Steve says... now hop on one leg Apple buyers = dumb,dumb,dumb

- Keep telling yourself that we (people who bought the iPhone on release day) are whiney and spoiled. We had the cash to get bragging rights... now we get an extra $100 in our pocket. How are we stupid again?

The following two Blogs came from an article a few days later about a woman who was suing Apple for $1,000,000.00!

- Stupid lawsuit...though she should have sued to get the phone unlocked with no scare tactics to disable by apple. . overrated ...company. apple

- who can i sue my house value went down this year???? the woman should use her iphone and get in touch with herself.

PROMPT 108

MORE BLOGGER'S GOG

These Bloggers were responding to an article about ATA Airlines detaining a passenger for using an iPhone while it was in 'airplane mode!'

✈ Great Entertainment: I love reading this stuff! Hey, we're all idiots to some extent--you and me included! I mean, you're voting for Hillary. Who's the bigger idiot? Probably GW...

✈ Flight Attendants Have Low IQs: You're dealing with people who are basically restaurant waitresses on a flying bus. What do you expect? Intelligence? Common sense? Next time you're in a restaurant getting lousy service, ask yourself: "If I was stuck in this restaurant for 3 to 4 hours, would I want them waiting on me?" Now imagine putting your life in their hands, or God forbid, that they worked for TSA.

✈ Another point: I think it may be possible that the ban on phones in planes may have more to do with the wireless carriers. A friend of mine, who has a pilot's license, once used his cell phone while flying solo years ago. He immediately received a call from his wireless carrier asking "WT_ did you do?" His phone pinged 150 towers during that one short call. I suppose if all passengers on all commercial flights were allowed to make calls while in the air it would bring the nation's cell networks to their knees.

PROMPT 109

✈ I used my iphone I even showed a flight attendant my phone not 3 weeks ago. Flying from Hilo to Oakland, no one cared. I don't think this is an ATA issue, but rather a stupid employee issue.

✈ ATA is ignorant: I'm a pilot and an electrical engineer; I say the ban on cell phones is unnecessarily conservative. The only possible equipment that could be remotely effected is the glide slope portion of the ILS receiver, since that is in a frequency band close to what the oldest analog cell phones used, but even then, it would have a low chance of interference. Most new cell phones use a completely different band and lower power. The ILS is only effective below about 4000 feet, so cell phones could be permitted above 10,000 feet. I still think they should be banned for the simple reason that people that blab on them can't keep their voices low and such people are worse then terrorists. As for "airplane mode" - well in that case, the device is no longer even a cell phone since the transciever is disabled, so there should be NO prohibition against iPhones in "airplane mode".

NOTE: CNET Networks is not responsible for the content of TalkBack posts submitted by our users.

MORE BLOGGERS' GOG

These Bloggers were responding to an article on Cyberchondriacs:

- I find more information on the web than my doctors have the time or inclination to share with me. Sometimes I discover things that help my doctor to better address my problems

- If it's not that bad I take the advice given but if it gets worse or is pretty bad to begin with I'll head into the doctors office.

- I am not a cyberchondriac but I diagnose my friends ailments all the time using the internet. Then they go to the doctor and I'm always right. I'm gonna start charging!

- Its like getting a second opinion!

- I am kind of a hypochondriac, but I never look up what I may (not) suffer from.

- Geek! Heal thy self!

- I gave up WebMd long ago, after it gave me a couple anxiety attacks. Everything could possibly mean death-headaches, being tired, having a runny nose. Gah.

PROMPT 111

ABOUT BLOGGING

When I was in netlingo.com, looking up the word, Blog, I came across a Link to an Article that peaked my interest. When I clicked on the Link, I found the Article, "Everything you Always Wanted to Know About Blogging (But Were Afraid to Ask)." However, I was too lazy to read the Article, but as I scrolled down, I found a Recap, which you will find on the following two pages!

A combination of the fact that it was midnight and the fact that I had been overwhelmed with the amount of 'STUFF' out there on the Information Highway, I found myself responding in my head to each of the ten points, laughing as I wrote each one down next to each of these points! I have put them in parenthesis after a hyphen! Having read them to several people, who laughed out loud, I decided to include them in this book!

PROMPT 112

Everything You Always Wanted to Know About Blogging
(But Were Afraid to Ask)*

(Well, before writing this book, the truth is, I never wanted to know anything about Blogging!!! And if I had wanted to know, I certainly wouldn't have been afraid to ask!)

1. Get a blog. - (Why would I want one and what would I do with it when I got it?)

2. Get a sitemeter (and a onestat). - (You want me to get how many more gadgets?............I don't think so!)

3. Leave comments. - (Somehow I don't think that you want to hear what I have to say!)

4. Use trackback. - (I prefer to go forward, not backwards......thank you very much!)

5. Take advantage of open posts - here, here, and here for example. - (Where, where and where.............for example?)

PROMPT 113

6. Find your tribe (memeorandum, technorati, etc). - (I didn't know I had a tribe......I wonder when I lost it! Maybe I should go out looking for it? Here I go awandering among..........!")

7. Email. - (Email who, what, where when.............wait, one more...........WHY?)

8. Carnivals. - (Is it in New Orleans, Rio de Janiero or Blogosphere, Blogistan, Blogspace or Blogland?)

9. Blogroll (and ping it too!). - (Can you eat it and play pong at the same time?)

10. Enter the Ecosystem. - (No way I'm going in there to interact with biological organisms in their physical environment! Are you totally crazy.........or what! Or maybe just a bit daft)!

*See the two links below to see what the expanded meanings are of the above 10 fundamentals every blogger needs to know...................if you must, of course!

*1. netlingo.com   2. Published online May 7, 2005 at The Mudville Gazette (reposted from 2005-04-28 21:12:08

PROMPT 114

'Technology Theme Party Menu'

Created by

patricia stutts

TECHNOLOGY THEME PARTY MENU

DROP DOWN MENU

Software Soup	Spicy Spyware Noodle Soup
Apple Appetizer	Mac Mini Salad with iCheesy Dressing
e-Entrée	Floppy Ribeye with Hard Drive Hollandaise
	Broadband Broccoli
	Stuffit Potatoes with Bacon Bytes
Backup Beverages	Blackberry Juice
	Out-Sourced Water
Digital Dessert	Chocolate Mouse Mousse with Clotted Anti-Virus Cream
After dinner cocktails	USB Port
	Downloaded Magnum of Champagne
Price of Meal	$.Froogle

PROMPT 117

PROMPT 118

TOP TEN LISTS

THINGS
TO
'HATE' AND 'LOVE'
ABOUT TECHNOLOGY!

THE TOP TEN LIST OF THINGS TO HATE ABOUT TECHNOLOGY

10. Accessories/Super Stores - Way too many choices!

9. Cables/plugs/USB ports/adapters - Couldn't they at least be Brand compatible.......honestly!

8. Information Highway - Gridlocked with way too much information! Not sure where to Enter or Exit!

7. Technology Era - It is outdated as soon as you take it out of the store! And before you learn how to use it............the next newest, latest and greatest model is out, which of course, you want!

6. Extended Warranties - Duped - We are already paying for a warranty that comes with the product. What does it say to the consumer when companies can't/won't stand behind their own workmanship!

5. Dropped cell phone calls!

4. Consumes more time than it saves! - 90% of the time I could have done it quicker manually!

3. It's a vacuum - sucks all of your money up (and like a vacuum you have to keep changing the bags (products) - way too often..............I might add!

2. Tech-Support – here, there and everywhere..............with all its Madness!

1. Prompts, dreaded prompts! Oh, to be a fly on the wall when their creators get prompted to death!

(Took me about 10 minutes to compile this list!)

PROMPT 121

THE TOP TEN LIST OF THINGS TO 'LOVE' ABOUT TECHNOLOGY

10.

9. GPS - when it works and you're not in a hurry to get where you're going!

8. Memory Sticks - backing up or transferring docs without having to take your computer with you.

7.

6. CD disks and DVD disks – nice and compact! Guess that's why they call them, Compact Discs!

5. Portable walk-about phones with hands-free speaker - better to multi-task with! Not tethered!

4. Medical equipment, (MRI, High-speed Dentist drills (Yikes), but still waiting for a woman to design mammogram equipment!

3.

2. Cell phones for emergencies/stuck in traffic - of course, they don't work in the Boonies where you may need them the most! But if you lose your life or are lost on a mountain, they can find you by your cell phone ping! Now how comforting is that.............well, I guess if you were lost it would be!

1. Computer – email, photos, videos, music (with less than 1 hour of tech-support a day!!)

(Took me about 7 days to attempt to compile this list...........sound familiar!)

PROMPT 122

MEET THE 'KEYBOARD HEAD'

HISTORY OF THE 'KeyBoard Head'

For weeks I searched for an Artist, trying to explain that I wanted them to create a 'Techno Character'; but not Flash Gordon and not the Transformers (nothing mean or evil looking, but rather a funky, quirky type character). A character that would eventually have a whole world of friends for future books.

Finally, after meeting students and teachers from local high schools, referrals, calling design companies and colleges, I had one creative soul who sent me something that was workable, but sadly she was unable to take on the project. However, the night before, out of total frustration and desperation, sitting at my desk, I picked up a pencil and started to doodle on a sheet of copy paper.

First I drew a square, then I slanted the bottom lines to make a chin. As I sat there staring at my Mac, I put the camera that is at the top of my laptop in the upper middle part of his face (the eye). Next I drew a rectangle for his mouth in the shape of a disk drive slot. He was pretty plain looking and as I was trying to decide what to do with the keyboard, it just seemed to take shape on top of his head. "Okay," I chuckled to myself, "He looks Funky/Quirky!" Then I added the sideburns! Again, don't know what possessed me but I started to dangle icons (i.e., @ sign, AOL, a Chad) from the keyboard on either side of his head. The Chad tickled my funny bone and I immediately erased all of the other icons and drew all Chads! The caption, "What's wrong? Haven't you ever seen dangling Chads before?" flittered across my mind................and I knew that I had my main character..........The Jester of the group; who can remove his keyboard and has a screen on his chest! The KeyBoard Head already has other characters that will show up in future books, which you will find on our Web Site (patafeabooks.com).

PROMPT 125

PROMPT 126

ABOUT THE AUTHOR

PATRICIA (Warren) STUTTS, a.k.a., patafea pecosa (ugly freckled duck), was born in New York City, but currently resides in Minnesota. At age six her family moved to Florida and at fourteen to Freeport, Grand Bahama. She started writing poetry as a child, but it wasn't until age seventeen that it became a serius hobby. Patricia's experiences and education include curriculum in Marine Science (tagged sharks for the U.S. Government in The Bahamas), scuba instruction (taught in The Bahamas), fencing (The Bahamian fencing Team/taught at U.O.P. in Stockton, CA), ESL Instructor (India/Chile), Butling & Household Management (MN/VA/NV), and Admin (Pentagon - JAG CORPS). She is the mother of two adult children who live in California. Motherhood, extensive travel and living abroad have added to her ability to keep a sense of humor with her at all times..........even in the face of reality!

Her travels include: Chile, Peru, Sweden, Germany, France, Taiwan, Hong Kong, Great Britain, Bali, Nepal, Holland, India and 44 States..............consequently, lots of books to be written!

ABOUT THE ILLUSTRATOR

Keila Ramos, (knachy© Productions) was born in Puerto Rico and moved to Minnesota to pursue an education in art. She graduated from the Minneapolis College of Art and Design (MCAD) with a BFA in Animation.

She currently teaches art, and also works as a freelancer. Her talent includes but is not limited to 2D, 3D and Stopmotion Animation, Anime, Film, Manga, Comics, Cartoons, Painting and Website Design. To her credit, Keila has worked for Nickelodeon, Tokyopop Books, Stonearch Books, SouthWest Journal, and many other clients.

Keila enjoys to party with her dog Brownie at home. Her hobbies are movies, videogames, drawing, and having a great time with friends. She loves her Mom, Brownie's mom (Cookie), and her friend Hector.

To see samples of her work, visit: www.knachy.com

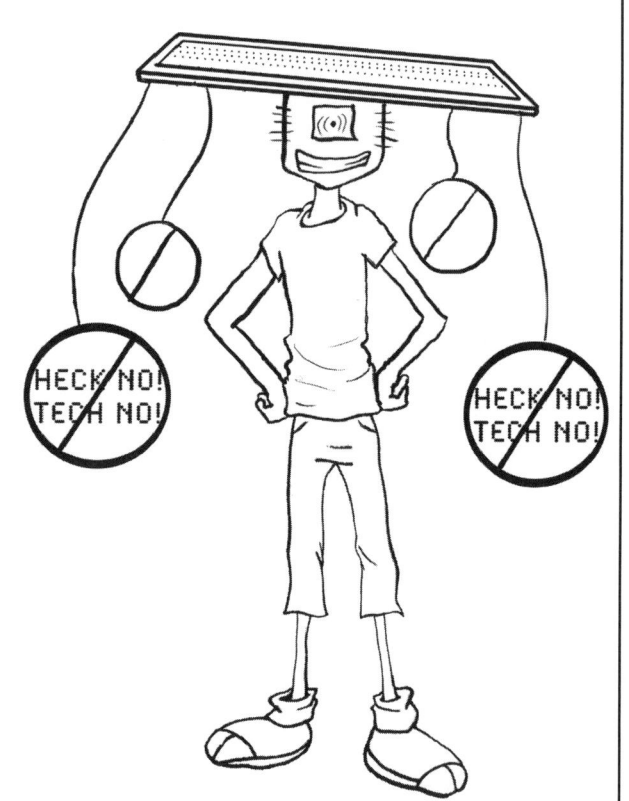

THANK YOU!

Thanks for reading, HECK NO! TECH NO! (A HUMOROUS GLIMPSE AT THE MADNESS OF TECHNOLOGY!)!

Hope you enjoyed reading it as much as I enjoyed writing it! Limericks, Poems and Haiku will be available as framed prints.

If you're a TECHNOCOWBOY, TECHNOHIPPIE or suffer from TECHNOHYSTERIA or TECHNOSIS, etc., you can have these illustrated pages hanging on your walls in your office or home!

There will also be T-Shirts available.

Great for gifts as well!

PROMPT 129

ORDER INFORMATION

- ORDER BOOKS

- PRINTS (LIMERICK/POEM & HAIKU PAGES AVAILABLE FRAMED)

- T-SHIRTS/BOOK BAGS

- SCHEDULE A BOOK SIGNING AND/OR READING FOR YOUR NEXT CONVENTION, LUNCHEON OR GROUP MEETING............FOR A HUMOROUS GOOD TIME!

CONTACT US AT:

~~patricia@patafeabooks.com~~
patafea@aol.com
www.patafeabooks.com

~~952-914-4324 (VoiceMail)~~

~~4737 County Road 101~~
~~Minnetonka, Minnesota 55345~~

patafea books ™

PROMPT 130

IT IS NOW SAFE TO CLOSE THIS BOOK!

IF YOUR BOOK WON'T CLOSE AUTOMATICALLY HOLD DOWN POWER BUTTON BELOW!

patafea books™